ASTRONOMY SCIENCE OF UNIVERSE

Thank you for choosing our Book

Disclaimer

This book contents is for informational and study purposes only. The publisher makes no claims, promises, or guarantees about the accuracy, completeness, or adequacy of the contents of this book and no legal liability or other responsibility is accepted by publisher for any errors, omissions, or statements on this book.

No part of this book may be reproduced or transmitted in any form or by any means, electronic or mechanical, including photocopying, recording or by any information storage and retrieval system, without written permission from the publisher.

About Book

Astronomy is the study of the science of universe including sun, moon, stars, planets, comets, gas, galaxies, gas, dust, cosmology and phenomenon. Astronomy and astrology both are historically related however astrology based on prediction and is no longer recognized as comprising anything to do with astronomy.

The book Astronomy Science of Universe has describing the observations of heavenly bodies. It is also focused to astrophysics. Astrophysics engaged to the study of the physics of astronomy and focuses on the behavior, properties and motion of objects in the space.

The book of astronomy offers quick bites into the most significant part of universe science.

Contents

Chapter 1: Introduction to Astronomy

Chapter 2: What is Cosmology?

Chapter 3: Astrophysics

Chapter 4: Galaxy

Chapter 5: Solar System

Chapter 6: Black Holes

Chapter 7: Nebula

Chapter 8: Supernova

Chapter 9: Dark Matter & Energy

Chapter 10: Stars

Chapter 11: The Moon

1

Introduction to Astronomy

Introduction

Astronomy is the science of universe over Earth's environment. The term is raised from the Greek origin astron for star, and nomos for preparation or rule. Astronomy is involved with astronomic matters and miracles – like planets, stars, galaxies and comets – as well as the important features of the Space, also called the Large Image. Extra precisely, astronomy is the learning of the source and growth of the Universe, the chemistry and physics of astronomic matters, and the estimation of their locations and gestures.

Astronomy signifies a world that is both grandiose and mysterious– a narrative regarding shining astronomic matters and the extent of the universe. Staring up at a twinkling night sky suggests outlandish and exclusive sensations, as if we have been specified a sight into the maximum essential secrecies of existence. The decoy of these universal mysteries was the stimulus that burned the appetite of lots of nowadays renowned astrophysicists.

Astronomy as well shows a much extra realistic character that is not closely as significant nowadays as it was in the previous. Since the period of our initial descendants, persons have utilized the gestures of astronomic matters to location themselves in place and period. Primitive persons considered the connection among the seasons and the distance of days to design their shooting and assembly actions. It was also by detecting the locations of the stars that the first agronomists determined when to establish and when to produce, and primary skippers can navigate the oceans blue.

In Astrography – the dimension of the locations of stars and planets – was the major profession of astrophysicists. There was a few amount of attention for astronomy from the common community too, since they supposed that the locations of astronomic matters prejudiced occasions that took position on Earth. Astrology, which is the skill of forecasting prospect proceedings depend on such explanations, was deliberated a division of astronomy and designed an essential portion of the astronomer's job for some centuries.

In the revival, progresses in arithmetic joined with the creation of new experimental tools provided increase to current astronomy. Studies into the power of gravity led to the formation of spiritual mechanism— a new division of astronomy that permissible the signals of astronomical substances to be arithmetically forecast for the first time forever. Astrometry and astronomic mechanism developed the two major arenas of learning for astrophysicists, while astrology was demoted to the rank of pseudo-science and no lengthier experienced by astrophysicists.

From the 19th century forwards, the finding of the electromagnetic range and the world of the particle urged on the growth of astrophysics, a new discipline in astronomy that is now deliberated to be the maximum significant.

Nowadays, Astronomy defend some disciplines—

- Astrometry, which is the accurate dimension of the location of planets and stars, and which is importantly simplified by the usage of modern computers and CCD cameras.
- Solar astronomy, which is the learning of the backgrounds and growth of stars.
- Galactic astronomy, which studies the construction and works of galaxies.
- Astrophysics, which studies the Cosmos and the physics of its components.
- Cosmology, which studies the source and development of the Cosmos as an entire.

Connected to these arenas of investigation are two corrections that lie extrain the area of biologists and geologists—global science, which is the learning of planets, astrobiology and comets, and a steroids, which studies the probability of life in the Cosmos.

Today, specialized astronomers have a sturdy contextual in astrophysics and their explanations are virtually continuously observed in an astrophysical setting. Some new philosophies are verified by explanations that authorize or disprove the offers or permit new thoughts to be progressive. The procedure can be supposed of as a repeated discussion among philosophy and attention. Unprofessional astronomers also play a significant role in investigation. The maximum grave laypersons vigorously contribute in the learning of moveable stars, the detection of new comets or asteroids, and other thrilling astrophysical effort.

The Past and Chronology

Astronomical actions have been experimental and chronicled since the beginning of evolution. Actions such as the increasing and location of the sun, the motion of stars in the sky, and astral and lunar eclipses arisen with such systematic designs that antique values cannot aid but notice and attempt to character what was occurring in the heavens. Lots of these explanations were intensely connected with traditional and spiritual principles.

The Ancient Greeks were mainly attentive in astronomy and were capable to estimate various exciting figures, such as the dimension of an expanse to the Sun. Arab, Chinese and Persian astrophysicist's maintained astronomy in the Central Ages, when slight development was created in Europe.

Through the Regeneration, an attention in astronomy was invigorated in Europe and several of the greatest significant terms initiate to display up in workbooks. Inventors like Newton, Kepler, Copernicus, and Galileo all existed and deliberate astronomy and physics about the 1600s. Hundreds of years later, the Space Age taken around a new age in astronomy, when immense telescopes can be created and even launched into path about the Earth to view matters further away than always previously.

Main detections are still created nowadays in the field of astronomy. The Hubble Space Telescope was launched in 1990 and remains to refer back data and magnificent pictures of galaxies and nebulae. As you can realize on display, here's visualize of the Supports of Formation occupied by the Hubble Space Telescope. This area is a brooder for new stars. The Hubble Telescope will finally have a successor. The James Webb Space Telescope is arranged to promotion in 2018.

Here is a summary history of significant astrophysical innovations—

- 30,000 BCE — bone models that demonstration the stages of the Moon. The initial accounts of primary astrophysical spectators.
- 700 BCE — astrologers account and forecast planetary conceals.

- 280 BCE — Aristarchus discovers the initial Sun-centered, or heliocentric, prototype of the galaxy as an alternate to the Earth-centered prototype.
- 164 BCE — initial account of Halley's Comet.
- 140 CE — Ptolemy prototypes an Earth-centric galaxy with spherical revolutions. This prototype correctly forecasts the locations of the planets, but as we recognize, Earth is not really the core of the galaxy.
- 1066 CE — the aspect of Halley's Comet outcomes in its presence in the Bayeux Tapestry.
- 420 CE — Ulugh Beg builds a laboratory in what is now Uzbekistan. He accumulates a star collection depend on his conceptions.
- 1543 CE — Copernicus distributes his prototype of the heliocentric galaxy.
- 1609 CE — Galileo utilizes a telescope for his conceptions.
- 1682 CE — Edmund Halley forecasts the reappearance of the shooting star, which is termed after him - Halley's Comet.
- 1783 CE — William Herschel defines the gesture of the Sun by universe.
- 1842 CE — Christian Johann Doppler defines what is now known as the Doppler Consequence or the alteration in wavelength of a signal if it travels near or gone from aspectator.

- 1916 CE — Albert Einstein defines his philosophy of common dependence, which must consequence in an increasing universe.
- 1929 CE — Edwin Hubble proves that the space is increasing.
- 1951 CE — Gerard Kuiper defines a belt of comets at the external border of the galaxy. This region is now known as the Kuiper Belt.

2

What is Cosmology?

Introduction

Cosmology is the discipline of science that studies the source and ultimate destiny of the universe. It is greatest nearly connected to the precise arenas of astrophysics and astronomy, however the latest century has as well carried cosmology nearly in track with key visions from atom physics.

In other language, we reach a captivating comprehension—our sympathetic of current cosmology derives from relating the actions of the biggest constructions in our universe (galaxy clusters, galaxies, stars, and planets) composed with those of the lowest constructions in our universe.

Astronomy Science of Universe

Past of Cosmology

The learning of cosmology is possibly one of the eldest procedures of hypothetical review into environment, and it initiated at several point in past when an antique person observed to the heavens. The antiques derived up with various rather good efforts to clarify these. Main between these in the western technical custom is the physics of the antique Greeks, who established a complete axial prototype of the universe which was sophisticated above the centuries until the period of Ptolemy, at which topic cosmology actually didn't grow more for some centuries, but in various of the information regarding the speeds of the several mechanisms of the structure.

The subsequent main development in this zone derived from Nicolas Copernicus in 1543, when he distributed his astronomy book on his deathbed, delineation the proof for his heliocentric prototype of the galaxy. The key insight that inspired this alteration in rational was the view that there was no actual cause to accept that the Earth comprises a basically advantaged location in the physical cosmos. This alteration in moulds is called the Copernican Value. Copernicus' heliocentric prototype converted even extra prevalent and acknowledged depend upon the work of Johannes Kepler, Tycho Brahe and Galileo Galilei, who collected considerable tentative proof in sustenance of the Copernican heliocentric prototype.

It was Sir Isaac Newton who was capable to carry all of these detections composed into really explanation the terrestrial waves, though. He had the instinct and insight to understand that the wave of matters dropping to the earth was like wave of matters looping the Earth. Since this wave was related, he understood it was possibly affected by the similar power, which known as gravity. By cautious reflection and the growth of a new arithmetic known as calculus and his three rules of wave, Newton was capable to make equations that defined this wave in a range of circumstances.

However Newton's law of gravity functioned at forecasting the wave of the heavens, there was one difficulty it was not precisely clear how it was functioning. The philosophy planned that matters with form attract each other across space, but Newton was not capable to grow a technical description for the device that gravity utilized to attain this. In arrange to clarify the mysterious, Newton relied on a common request to God - fundamentally, matters perform this mode in reply to God's perfect occurrence in the universe. To acquire a physical clarification would wait above two centuries, until the appearance of virtuosity whose intelligence can conceal even that of Newton.

Current Cosmology

Common irrelativeness and the big bang

Newton's cosmology governed science until the initial 20 century, when Albert Einstein established his philosophy of common irrelativeness, which delimited the technical sympathetic of gravity. In Einstein's new construction, gravity was affected by the winding of 4-dimensional space-time in reaction to the occurrence of an enormous entity, such as a star, a galaxy, or even a planet.

One of the exciting insinuations of this new construction was that space-time itself was not in balance. In equally small order, researchers understood that common irrelativeness forecast that space-time would both enlarge and bond. Consider Einstein supposed that the universe was really eternal, he presented a cosmological continuous into the philosophy, which given a weight that countered the growth or reduction. Though, when astronomer Edwin Hubble finally exposed that the universe was in effect increasing, Einstein understood that he would create an error and detached the cosmological continuous from the philosophy.

If the universe was increasing, then the usual deduction is that if you were to reverse the universe, you would realize that it should have initiated in a small, thick cluster of substance. This philosophy of how the universe initiated developed known as the big bang philosophy. This was a provocative philosophy by the central periods of the 20 century, as it competed for supremacy beside Fred Hoyle's stable state philosophy. The detection of the cosmic microwave contextual contamination in 1965, though, inveterate a forecast that had been created concerning the big bang, so it converted broadly established between physicists.

Although he was verified incorrect regarding the stable state philosophy, Hoyle is attributed with the main growths in the philosophy of stellar nucleon synthesis, which is the philosophy that hydrogen and other light particles are converted into weightier particles in the nuclear containers known as stars, and spittle out into the universe upon the star's expiry. These weightier particles then go on to form into water, planets, and eventually lifecycle on Earth, with persons! Therefore, in the language of several impressed cosmologists, we are all made from stardust.

Anyhow, rear to the development of the universe. As researchers increased extra knowledge regarding the universe and extra cautiously dignified the cosmic microwave contextual energy, there was a difficulty. As comprehensive dimensions were occupied of astrophysical data, it developed clear that conceptions from substantial physics required to play a tougher part in sympathetic the initial stages and development of the universe. This arena of hypothetical cosmology, however still greatly hypothetical, has developed rather lush and is sometimes known as substantial cosmology.

Substantial physics disclosed a universe that was attractive near to being constant in power and substance, but was not totally constant. Though, any variations in the primary universe would have extended highly above the billions of years that the universe extended and the variations were greatly minor than one would suppose, so cosmologists had to figure out a mode to clarify a dissimilar primary universe, but one which had simply very minor variations.

Come Alan Guth, anatomy physicist who undertook this difficulty in 1980 with the growth of rise philosophy. The variations in the primary universe were small substantial variations, but they fast extended in the primary universe because of an ultra-fast period of extension. Astronomical explanations since 1980 have maintained the forecasts of the rise philosophy and it is now the agreement lookout between maximum cosmologists.

Secrecies of Current Cosmology

Although cosmology has progressive greatly above the latest century, there are still some open secrecies. Actually, two of the middle secrecies in current physics are the leading difficulties in astrophysics and cosmology—

- Dark substances — several galaxies are moving in a mode that cannot be completely clarified depend on the quantity of substance that is experiential in them, but which can be clarified if there is additional hidden substance in the galaxy. This additional substance - which is forecast to start around 25% of the universe, depend on most current dimensions - is known as dark substance. As well as astronomical explanations, experimentations on Earth such as the Cryogenic Dark Substance Exploration are attempting to straight perceive dark substance.
- Dark power — in 1998, astronomers tried to identify the rate at which the universe was reducing down ... but they establish that it was not reducing down. Actually, the quickening rate was moving up. It appears that Einstein's cosmological continuous was required after all, but in its place of allotment the universe as a

state of balance it really appears to be aggressive the galaxies separately at a sooner and sooner rate as time goes on. It's unidentified precisely what is causing this disgusting gravity, but the name physicists have specified to that matter is dark power. Astronomical explanations forecast that this dark power creates up around 70% of the universe's matter.

There are several other recommendations to clarify these uncommon outcomes, such as Altered Newtonian Subtleties and adjustable speed of light cosmology, but these substitutes are measured peripheral philosophies that are not recognised between lots of physicists in the arena.

Background of the Universe

It is value noticing that the big bang philosophy really defines the mode the universe has changed since soon after its formation, but cannot provide some direct material regarding the real backgrounds of the universe.

This is not to declare that physics can express us nothing regarding the backgrounds of the universe, when physicists discover the minimum balance of space, they discovery that substantial physics outcomes in the formation of simulated atoms, as demonstrated by the Casimir result. Actually, rise philosophy forecasts that in the nonappearance of some substance or power, and then space-time would enlarge. Occupied at expression value, this thus provides researchers a sensible clarification for how the universe can primarily arise into being. If there were a factual nothing - no substance, no power, no space-time then that nothing would be uneven and would initiate producing substance, power, and an increasing space-time.

3

Astrophysics

Introduction

Astrophysics is accurately space science. Precisely, it is a division of space science that put on the laws of chemistry and physics in a try to recognize the maximum enormous substances in the universe. Clearly, astrophysics is regarding extra than only stars. It's regarding sympathetic galaxies, nebulae, black holes, planets, and all of the further substances nomadic in the cosmos. Since these substances openly prejudiced our development, sympathetic their precise devices are of the greatest importation.

There are two twigs of this discipline —Astronomy and Cosmology.

Cosmology is the learning of the cosmos at big. For instance, a cosmologist can learn the source, development, and ultimate destiny of the universe. Astronomy is the learning of separate substances or constructions. Astrophysics is the arena that makes physical philosophies for the minor to average-size constructions in the universe. If it noises like they all merger collected a bit, it's because they organize.

Astrophysics is a division of space science that smears the rules of chemistry and physics to clarify the birth, lifespan and expiry of stars, galaxies, planets, nebulae and further substances in the universe. It has two fraternal sciences, cosmology and astronomy, and the strokes among them distortion.

In the maximum inflexible logic—

- Astronomy estimates locations, radiances, waves and other features.
- Astrophysics makes physical philosophies of minor to average-size constructions in the universe.
- Cosmology does this for the biggest constructions, and the universe as a complete.

In exercise, the three occupations form an integrated family. Request for the location of a nebula or what type of light it releases, and the astrophysicist can response first. Request what the nebula is created of and how it designed and the astrophysicist will speak up. Request how the data proper with the creation of the universe, and the cosmologist would possibly jump in.

Aims of Astrophysics

Astrophysicists try to find to recognize the universe and our position in it. At the National Aeronautics and Space Administration, the aims of astrophysics are to determine how the universe functions, discover how it initiated and progressed, and exploration for existence on planets nearby other stars, conferring NASA's website.

It Initiated With Newton

Whereas astronomy is one of the eldest sciences, hypothetical astrophysics initiated with Isaac Newton. Previous to Newton, astrophysicists defined the waves of wonderful bodies by composite numerical representations deprived of a physical base. Newton displayed that an only philosophy instantaneously clarifies the paths of planets and moons in space and the route of a projectile on Earth, this additional to the body of signal for the surprising assumption that the heavens and Earth are topic to the similar physical laws. Maybe what maximum totally divided Newton's prototypical from earlier ones is that it is analytical with evocative. Depend on deviations in the Newtonian path of Uranus, astrophysicists forecast the location of a new planet, which was then experiential and called Neptune. Being extra polative with evocative is the symbol of a matured science, and astrophysics is in this classification.

Indicators in Astrophysics

As the only mode we interrelate with removed matters is by detecting the emission they produce, much of astrophysics has to do with assuming philosophies that clarify the apparatuses that generates this radioactivity, and offer thoughts for how to remove the maximum data from it. The leading thoughts regarding the nature of stars arisen in the mid-19th century from the prospering science of spectral study, which denotes perceiving the precise rates of light that specific matters engage and produce when animated, spectral study residues necessary to the three some of space sciences, both supervisory and analysis new philosophies. Primary spectroscopy given the main signal that stars comprise matters also extant on Earth. Spectroscopy exposed that various nebulae are virtuously vaporous, while various comprise stars. This late assisted adhesive the knowledge that various nebulae were not nebulae at all — they were further galaxies!

In the initially 1920s, Cecilia Payne exposed, by spectroscopy, that stars are mainly hydrogen, the ranges of stars also permitted astrophysicists to define the speed at which they passage near or gone from Earth. Just like the rigorous an automobile produces is dissimilar affecting nears us or gone from us, as of the Doppler move, the ranges of stars will variation in the similar mode, in the 1930s, by merging the Doppler move and Einstein's philosophy of common irrelativeness, Edwin Hubble given hard indication that the universe is increasing. This is also forecast by Einstein's philosophy, and collected from the fundamental of the Big Bang Philosophy.

Also in the mid-19th century, the physicists Gustav Von Helmholtz and Lord Kelvin gambled that gravitational failure can power the sun, but finally understood that power created this mode would only last 100,000 years. After Fifty years, Einstein's well-known E=mc2 equation provided astrophysicists the main evidence to what the factual basis of power can be. As nuclear physics, substantial mechanism and atom physics produced in the first partial of the 20th century, it converted probable to express philosophies for how nuclear synthesis can influence stars. These philosophies define how stars procedure, alive and expire, and effectively clarify the experimental supply of kinds of stars, their ranges, radiances, ages, and other characteristics.

Astrophysics is the physics of stars and other aloof forms in the universe, but it also successes near to home. Giving to the Big Bang Philosophy, the main stars were virtually completely hydrogen. The nuclear synthesis procedure that strengthens those crashes collected hydrogen particles to form the weightier component helium. In 1957, the husband-and-wife astrophysicist group of Margaret Burbidge and Geoffrey, accompanied by physicists Fred Hoyleand William Alfred Fowler, displayed how, as stars age, they generate weightier and weightier components, which they permit on to future generations of stars in forever-bigger amounts. It is just in the last periods of exists of extra new stars that the components creating up the Earth, such as oxygen (30.1 percent), silicon (15.1 percent), iron (32.1

percent), are formed. Additional of these components is carbon, which collected with oxygen, framework the unpackaged of the form of all existing objects with us. Therefore, astrophysics says us that, though we are not completely stars, we are completely stardust.

How Astrophysicists Perform?

Briefly, astrophysicists try to recognize the universe and our location in it. At NASA, the aims of astrophysics are to determine how the universe functions, discover how it initiated and progressed, and exploration for lifespan on planets nearby other stars. And every year, certainly, every day carries us a small bit nearer to sympathetic the complete creation of the cosmos. New and improved machineries permit us to aristocrat beyond back into the past of the universe and sight constructions that are billions of years old, carrying us nearer and nearer to the instant of the big bang.

In a try to response these queries we have journeyed to the moon, flied historical all the main planets in our area, and even left the galaxy. We have produced the Hubble Space Telescope and the Worldwide Space Station in an effort to acquire nearer to the cosmos. These combined exertions not only rise our sympathetic of the universe, they aid to make bonds among the many countries of the world. And these combined exertion said us construct the prospect. Expertise formed for space examination has direct to cars, planes, and safer homes. The growths produced for the ISS have permissible us to extra efficiently warmth our homes, pure our water, and nourish Earth's ever growing population.

4

Galaxy

Introduction

We live on a planet known as Earth which is the portion of our solar system. It is a little part of the Cloudy Way Galaxy. A galaxy is an enormous group of dust, gas, and billions of stars and their solar systems. A galaxy is detained collected by gravity. Our galaxy, the Cloudy Way, also has a super massive black hole in the central.

When you see up at stars in the night sky, you are observing other stars in the Cloudy Way. If it's actually dark, distant away from lights from towns and houses, you can even observe the dusty groups of the Cloudy Way give across the sky. There are lots of galaxies further ours, although. There are so a lot of, we can't even computation them all yet. The Hubble Space Telescope observed at a minor cover of space for 12 days and establishes 10,000 galaxies, of all dimensions, forms, and colors. Some researchers consider there can be as lots of as one hundred billion galaxies in the universe.

Sometimes galaxies acquire too near and crash into all other. Our Cloudy Way galaxy will sometime bump into Andromeda, our nearby immense neighbor. It won't occur for regarding five billion years. But even if it occurred tomorrow, you may not notice. Galaxies are so large and banquet out at the ends that even although galaxies bump into all other, the solar systems and planets frequently don't get near to bumping.

How Galaxies Born?

Galaxies are made of dust, stars and dark substance, all detained collected by gravity. They derive in a range of forms, dimensions and ages, and lots of have black holes at their cores. Galaxies comprise a dissimilar amount of star clusters, star systems, planets and kinds of stellar clouds. In among them is a thin stellar average of gas, dust and inter stellar emissions. The black holes at the core of maximum galaxies are deliberated to be the main driver of vigorous immense centers establish at the central, and their environs creates massive quantities of power that astrophysicists can see above excessive reserves. Substantial nearby the black hole is augmented away by its jets. Other galaxies comprise substances like quasars, the most active forms in the universe, at their centers.

Galaxies are characterized giving to their specious form, assigned to as their graphic morphology. A general appearance is the oval galaxy which has an oval-formed light outline. Twisting galaxies are disk-formed with dirty, twisted supports, and those with unequal forms are called unequal galaxies and naturally create from disturbance by the gravitational pull of adjoining galaxies. Connections among adjoining galaxies, which can result in a fusion, sometimes encourage considerably enlarged events of star creation important to starburst galaxies.

Categories of Galaxies

The maximum extensively utilized arrangement system for galaxies is depend on one invented by Edwin P. Hubble and other sophisticated by astrophysicist Gerard de Vaucouleurs. It utilizes the three major kinds, and then other breakdowns them down by precise features. In this age of multi-wavelength detecting, the sub-categorizations also comprise indicators for such features as a galaxy's star-creation rate and age range of its stars.

Twist Galaxies

Twist galaxies are the maximum general kind in the universe. Our Cloudy Way is a curved, as is the quite near-by Andromeda Galaxy. Twists are big spinning disks of nebulae and stars, within an explosive of dark substance. The essential bright area at the central of a galaxy is known as the enormous bulge. Lots of twists have a corona of stars and star collections arranged overhead and under the disk.

Twists that have big, bright bars of stars and substantial wounding across their dominant units are known as striped twists. A big mainstream of galaxies have these bars, and astrophysicists learning them to recognize what role they play in the galaxy. As well as bars, lots of twists can also comprise super massive black holes in their centers. Subcategories of twists are definite by the features of their bulges, twisting supports, and how strongly those spiral supports are.

Oval Galaxies

Oval galaxies are unevenly egg-formed originate mostly in galaxy bunches and lesser dense collections. Most oval comprise elder, short-form stars, and as they deficiency a great contract of star-creating air and dirt clouds, there is small new star creation arising in them. Oval can have as limited as a hundred million to maybe a hundred trillion stars, and they can variety in mass from a limited thousand light-years across to extra than a limited hundred thousand. Astrophysicists now suspicious that all oval has a dominant super massive black hole that is connected to the form of the galaxy them self. There are several subcategories of oval, with dwarf oval with properties that place them anywhere among even oval and the strongly join collections of stars known as spherical clusters.

Unequal Galaxies

- Unequal galaxies are as their name recommends —unequal in form. The finest instance of an unequal that can be understood from Earth is the Minor Magellanic Mist. Unequal generally do not have sufficient construction to characterize them as coils or oval. They can demonstration various bar construction, they can have vigorous areas of star creation, and several lesser ones are listed as dwarf unequal, extremely like the extremely initial galaxies that designed around 13.5 billion years before. Unequal are categorized by their constructions.

5

Solar System

Introduction

The Solar System is the gravitationally certain system including the Sun and the substances that path it, both directly and indirectly. Of those substances that path the Sun directly, the biggest eight are the planets, with the residue being expressively minor substances, such as dwarf planets and minor Solar System forms. Of the substances that path the Sun indirectly, the moons, two are bigger than the lowest planet, Mercury. The massive mainstream of the system's form is in the Sun, with maximum of the residual form limited in Jupiter. The four minor internal planets, Mercury, Venus, Earth and Mars, are earthly planets, being mainly collected of metal and rock. The four external planets are massive planets, being considerably extra gigantic than the earthly. The two biggest, Saturn and Jupiter are gas hulks, being collected mostly of helium and hydrogen; the two outmost planets, Uranus and Neptune, are snow hulks, being collected mainly of matters with comparatively great melting points associated with helium and hydrogen, known as snows, such as methane, water and ammonia. All planets have virtually round paths that lie in a closely even disc known as the ecliptic.

The Solar System also comprises minor substances. The asteroid belt, which lies among the paths of Jupiter and Mars, mainly comprises substances collected, like the earthly planets, of metal and rock. Over Neptune's path lie the Kuiper belt and dispersed disc, which are inhabitants of trans-Neptunian substances collected mainly of snows and over them a recently exposed inhabitants of sednoids. In these inhabitants are some dozen to probably tens of thousands of substances big sufficient that they have been round by their individual gravity. Such substances are characterized as dwarf planets. The solar airstream, a stream of exciting atoms graceful away from the Sun, makes a bubble-like area in inter planetary middle called the heliosphere. The heliosphere is the point at which weight from the solar airstream is equivalent to the contrasting weight of inters planetary airstream; it spreads out to the border of the dispersed disc. The Oort mist, which is supposed to be the foundation for lengthy-period comets, can also occur at aloofness unevenly a thousand times more than the heliosphere. The Solar System is situated in the Orion Arm, 26,000 light-years from the middle of the Cloudy Way.

The Sun

The Sun is the Solar System's star and by faraway its maximum enormous element. Its big form generates heats and thicknesses in its central great sufficient to sustain atomic synthesis of hydrogen into helium, creating it a major-arrangement star. This discharges a massive quantity of power, largely emitted into space as electromagnetic radioactivity climaxing in observable light.

The Sun is a G2-kind major-order star. Warmer major-order stars are extra shining. The Sun's heat is intermediate among that of the warmest stars and that of the coolest stars. Stars sunnier and warmer than the Sun are infrequent, while considerably regulator and chiller stars, called red dwarfs, make up 85% of the stars in the Cloudy Way.

Planets

Beneath is a passing indication of the eight main planets in our solar system, in arrange from the internal solar system external—

Mercury

The nearby planet to the sun, Mercury is just a bit bigger than Earth's moon. Its day side is burned through the sun and can influence 840 grades Fahrenheit, but on the night side, heats descent to hundreds of grades under cold. Mercury has effectively no environment to engross fireball influences, so its exterior is pocked with hollows, just like the moon. Above its four-year task, NASA's envoy Spacelab has discovered opinions of the planet that have defied astrophysicists' opportunities.

- Invention — recognized to the antiques and observable to the nude eye.
- Called for — Envoy of the Roman gods.
- Width — 3,031 miles (4,878 km).
- Path — 88 Earth days.
- Day — 58.6 Earth days.

Venus

The second planet from the sun, Venus is offensively warm, even warmer than Mercury. The environment is poisonous. Researchers define Venus' condition as a blockbusting conservatory consequence. Its dimension and construction are parallel to Earth, Venus' dense, poisonous environment ruses warmth in a blockbusting conservatory consequence, strangely, Venus rotations gradually in the conflicting way of maximum planets.

The Greeks supposed Venus was two dissimilar substances — one in the morning sky and additional in the evening. As it is frequently sunnier than any other entity in the sky — excepting for the moon and sun — Venus has produced lots of unidentified flying object reports.

- Invention — recognized to the antiques and observable to the nude eye.

- Called for — Roman goddess of beauty and love.
- Width — 7,521 miles (12,104 km).
- Path — 225 Earth days.
- Day — 241 Earth days.

Earth

Earth is a water world, with two-thirds of the planet enclosed by sea, the third planet from the sun. It's the just world recognized to port life. Earth's environment is ironic in life-supporting oxygen and nitrogen. Earth's area revolves regarding its axis at 1,532 feet per second—somewhat extra than 1,000 mph— at the equator, the planet vitalities nearby the sun at extra than 18 miles per second.

- Width — 7,926 miles (12,760 km).
- Path — 365.24 days.
- Day — 23 hours, 56 minutes.

Mars

The fourth planet from the sun, is an icy, dirty place. The dust, an iron oxide, provides the planet its ruddy cast. Mars parts resemblances with Earth: It is stony, has valleys and mountains, and tempest structures oscillating from restricted tornado-like dirt sprites to planet-overwhelming dust tempests. It ices on Defaces. And Mars docks water snow. Researchers consider it was one time warm and wet, by nowadays it's icy and desert-like.

Mars environment is too slim for fluid water to be on the superficial for any distance of time. Researchers consider antique Mars would have had the circumstances to sustenance exist, and there is courage that symbols of previous life — perhaps even current biology — can be on the Red Planet.

- Invention — recognized to the antiques and observable to the nude eye.
- Called for — Roman god of conflict.
- Width — 4,217 miles (6,787 km).
- Path — 687 Earth days.
- Day — just extra than one Earth day (24 hours, 37 minutes).

Jupiter

The fifth planet from the sun, Jupiter is enormous and is the maximum huge planet in our solar system. It's a mostly vaporous biosphere, mostly helium and hydrogen. Its whirling clouds are interesting because of dissimilar kinds of dash vapors. A large character is the Great Red Spot, a huge squall which has stormed for hundreds of years. Jupiter has a sturdy attractive arena, and with lots of moons, it appearances a minute similar a small solar system.

- Invention — recognized to the antiques and observable to the nude eye.
- Called for — Monarch of the Roman gods.
- Width — 86,881 miles (139,822 km).
- Path — 11.9 Earth years.
- Day — 9.8 Earth hours.

Saturn

The sixth planet from the sun is recognized maximum for its circles. When Galileo Galilei main deliberate Saturn in the primary 1600s, he supposed it was a thing with three portions. Not expressive he was realizing a planet with circles, the bewildered astrophysicist arrived a little illustration — a sign with one big ring and two lesser ones — in his notebook, as a noun in a verdict telling his innovation. More than 40 years later, Christian Huygens recommended that they were rings. The rings are generated of ice and support. Investigators are not yet convinced how they intended. The steamy planet is naturally hydrogen and helium. It has lots of moons.

- Invention — recognized to the antiques and detectible to the nude eye.
- Called for — Roman god of farming.
- Width — 74,900 miles (120,500 km).
- Path — 29.5 Earth years.
- Day — Around 10.5 Earth hours.

Uranus

The seventh planet from the sun, Uranus is an eccentric. It's the only massive planet whose orbit is closely at correct viewpoints to its path — it fundamentally paths on its cross. Astronomers consider the planet struck with various other planet-dimension objectives extensive previously, causing the slope. The slope reasons risky periods that last 20-plus years and the sun strokes down on one opposite or the other for 84 Earth-years. Uranus is regarding the similar dimension as Neptune. Methane in the environment provides Uranus its blue-green shade. It has many moons and palecircles.

- Invention — 1781 by William Herschel.
- Called for — Exemplification of heaven in antique legend.
- Width — 31,763 miles (51,120 km).
- Path — 84 Earth years.
- Day — 18 Earth hours.

Neptune

The eighth planet from the sun, Neptune is identified for sturdy airstreams — sometimes quicker than the speed of noise. Neptune is distant out and icy. The planet is extra than 30 times as distant from the sun as Earth. It has a stony central. Neptune was the primary planet to be forecast to occur through with mathematics, previously it was discovered. In discretions in the path of Uranus led French astrophysicist Alexis Bouvard to recommend several further can be applying a gravitational tug. German astrophysicist Johann Galle utilized computations to aid discover Neptune in a telescope. Neptune is around 17 times as enormous as Earth.

- Invention — 1846.
- Called for — Roman god of marine.
- Width — 30,775 miles (49,530 km).
- Path — 165 Earth years.
- Day — 19 Earth hours.

Pluto

Once the ninth planet from the sun, Pluto is different further planets in lots of compliments, it is lesser than Earth's moon. Its path transmits it within the path of Neptune and then mode out over that path. From 1979 until early 1999, Pluto had really been the eighth planet from the sun. Then, on Feb. 11, 1999, it overlapped Neptune's track and once more converted the solar system's most aloof planet — until it was relegated to dwarf planet position. Pluto will remain over Neptune for 228 years. Pluto's path is slanted to the major even of the solar system — where the further planets path — by 17.1 grades. It's an icy, stony ecosphere with only an extremely transient environment. NASA's New Prospects task completed past's first flyby of the Pluto system on July 14, 2015.

- Invention — 1930 by Clyde Tombaugh.
- Called for — Roman god of the gangland, Hells.

- Width — 1,430 miles (2,301 km).
- Path — 248 Earth years.
- Day — 6.4 Earth day.

Comets

Comets are minor Solar System groups, naturally just a small number of kilometers across, collected mainly of unstable snows. They have extremely unconventional tracks, usually a perihelion in the tracks of the internal planets and an aphelion distant over Pluto. When a comet arrives the internal Solar System, its nearness to the Sun reasons its frozen area to transfer and ionize, generating a coma— a lengthy end of dust and gas frequently observable to the nude eye.

Small-time comets have tracks durable fewer than two hundred years. Extended-time comets have tracks durable thousands of years. Small-time comets are supposed to create in the Kuiper belt, while extended-time comets, such as Hale–Bopp, are supposed to create in the Oort cloud. Lots of comet collections, such as the Kreutz Sungrazers, designed from the disintegration of sole parental. Various comets with hyperbolic tracks can create outdoor the Solar System, but defining their exact tracks is problematic. Old comets that have had maximum of their unstable ambitious out by solar heating are frequently characterized as asteroids.

Asteroids

Asteroids are minor, stuffy stony ecospheres rotating nearby the sun that are too minor to be known as planets. They are also called small planets or planetoids. In whole, the form of all the asteroids is fewer than that of Earth's moon. But not withstanding their size, asteroids can be risky. Lots of have hit Earth in the past, and extra will smash into our planet in the prospect. That is one cause researchers learning asteroids and are excited to study additional regarding their numbers, tracks and physical features. If an asteroid is controlled our approach, we want to recognize that.

Most asteroids lie in a massive circle among the tracks of Jupiter and Mars. This major asteroid strap clutches extra than 200 asteroids bigger than 60 miles in width. Researcher's evaluation the asteroid belt also comprises more than 750,000 asteroids larger than three-fifths of a mile in width and millions of minor ones. Not all in the major belt is an asteroid — for example, comets have currently been exposed there, and Ceres, once supposed of just as an asteroid, is now also deliberated a dwarf planet.

Lots of asteroids lie external the major belt, for example, a figure of asteroids known as Trojans lie between Jupiter's detour tracks. Three clusters — Amors, Apollos, and Atens — called close-Earth asteroids track in the internal solar system and sometimes cross the track of Earth and Mars.

6

Black Holes

Introduction

Black hole is everything but blank space. Quite, it is an excessive quantity of substance filled into an extremely minor region - consider of a star ten times extra enormous than the Sun embraced into a range about the width of New York City. The outcome is a gravitational arena so durable that nothing, not even light leakage. In current years, NASA devices have coated a new image of these odd substances that are, to lots of, the maximum captivating substances in space.

Though the word was not invented until 1967 by Princeton physicist John Wheeler, the impression of a thing in space so enormous and thick that light cannot leakage it has been round for centuries. Maximum superbly, black holes were forecast by Einstein's philosophy of common irrelativeness, which displayed that when an enormous star expires, it leaves after a minor, thick remainder central. If the centre's mass is extra than around three times the mass of the Sun, the equations displayed, the power of gravity overcomes all other powers and generates a black hole.

Researchers cannot directly detect black holes with telescopes that perceive x-rays, light, or other procedures of electromagnetic emission. We can, though, conclude the occurrence of black holes and learning them by perceiving their consequence on other substance close. If a black hole permits by a cloud of stellar substance, for instance, it will draw substance inner in a procedure called accretion. A same procedure can arise if a usual star permits near to a black hole. In this case, the black hole can tear the star separately as it pulls it to them self, as the involved substance quickens and temperatures up, it produces x-rays that emit into space. Modern innovations give several enticing proof that black holes have a affected effect on the areas nearby them – producing influential gamma ray spurts, consuming near stars, and encouraging the development of new stars in some regions while obstructionist it in others.

End of Star is Black Hole Opening

Maximum black holes start from the leftovers of a big star that expires in a supernova explosion. If the complete form of the star is big sufficient, it can be established hypothetically that no power can preserve the star from failing below the effect of gravity. Though, as the star failures, an odd object arises. As the exterior of the star approaches a fantasy exterior known as the occurrence prospect, time on the star reduces comparative to the time reserved by viewers distant away. When the exterior spreads the occurrence prospect, time positions still, and the star can failure no more - it is a cold failing thing.

Incredible Facts of Black Holes

Visualize substance filled so thickly that nothing leakage. Not a light, not a moon and not even planet. That's what black holes are — a spot where gravity's pull is enormous, finish up being risky for anything that unintentionally waifs by. Beneath we have 10 facts regarding black holes — only some tit bits 'regarding these captivating substances.

Fact 1: You can't normally view a black hole.

As a black hole is really black — no light can leakage from it — it's unbearable for us to feel the gap directly by our devices, no problem with the type of electromagnetic emission you utilize. The basic is to appearance at the black holes consequences on the close atmosphere, points out NASA. Speak a star occurs to acquire too near to the black hole, for instance. The black hole usually pulls on the star and rips it to scraps. When the problem from the star initiates to drain near the black hole, it acquires quicker, acquires warmer and radiances radiantly in X-rays.

Fact 2: Appearance out, our cloudy way probable has a black hole.

Astrophysicists declare, though there is perhaps an enormous super massive black hole prowling in the central of our galaxy. Fortunately, we are nowhere close this enormous — we are around two-thirds of the mode out from the center, comparative to the respite of our galaxy — but we can surely perceive its consequences from far. For instance: the European Space Agency declares it's four million times extra enormous than our Sun, and that it's enclosed by amazingly burning gas.

Fact 3: Disappearing stars make astral black holes.

About you have a star that's around 20 times extra enormous than the Sun. Our Sun is going to end its life silently; when its atomic petroleum burns out, it'll sluggishly disappear into a white dwarf. That is not the situation for distant extra enormous stars. When those giants run out of petroleum, gravity will overpower the usual weight the star preserves to stay its form constant. When the weight from atomic reactions downfalls, giving to the Space Telescope Science Institute, gravity aggressively over powers and downfalls the central and further coatings are threw into space. This is known as a supernova, the residual central downfalls into a distinctiveness - – an advert of immeasurable thickness and virtually no capacity. That is additional term for a black hole.

Fact 4: Black holes derive in a range of dimensions.

There are as a minimum three kinds of black holes, NASA speaks, ranging from comparative near misses to those that govern a galaxy's center. Primordial black holes are the lowest types, and variety in dimension from one particle's dimension to a mountain's form. Astral black holes, the most general kind, are up to 20 times extra enormous than our personal Sun and are probable scattered in the dozens in the Cloudy Way. And then there are the huge ones in the centers of galaxies, known as super massive black holes. They are all extra than one million times extra enormous than the Sun. How these creatures designed is still being observed.

Fact 5: Strange time material occurs nearby black holes.

This is best demonstrated by one individual dropping into a black hole although a different person observes. From Lucky's viewpoint, Unluckiest time clock seems to be marking sluggish and sluggish. This is in accord with Einstein's philosophy of common irrelativeness, which speaks that time is pretentious by how quick you go, when you are at risky rapidity close to bright. The black hole twists space and time so much that Unluckiest time seems to be running sluggish. From Unluckiest viewpoint, though, their clock is running usually and Lucky's is running quick.

Fact 6: The first black hole was not exposed until X-ray astronomy was utilized.

Cygnus X-1 was first initiate in inflatable tours in the 1960s, but was not recognized as a black hole for regarding alternative period. Rendering to NASA, the black hole is 10 times extra enormous to the Sun. Adjacent is a blue super giant star that is regarding 20 times extra enormous than the Sun, which is flow because of the black hole and making X-ray radiations.

Fact 7: The adjoining black hole is probable not 1,600 light-years gone.

An in accurate dimension of V4641 Sagittarius led to a slide of broadcast reports some year's rear saying that the adjoining black hole to Earth is astonishingly near, just 1,600 light-years gone. Not nearby sufficient to be deliberated risky, but mode nearer than supposed. More study, though, demonstrations that the black hole is possible more away than that. Observing at the revolution of its acquaintance star, between further features, produced a 2014 outcome of extra than 20,000 light years.

Fact 8: We are not undisputable if wormholes occur.

Prevalent science-narrative subject anxieties what occurs if somebody drops into a black hole. Several persons consider these substances are a kind of wormhole to other portions of the Universe, than-bright tourism probable. But as this Smith Sonian Magazine object points out, everything is probable since we still have a lot to digit out regarding physics. Since we do not yet have a philosophy that dependably merges common irrelativeness with substantial mechanism, we do not distinguish of the complete zoo of probable space-time constructions that can quarter wormholes, said Abi Loeb, who is with the Harvard-Smithsonian Center for Astrophysics.

Fact 9: Black holes can be risky if you acquire too near.

Similar persons behind a birdcage, it is okay to detect a black hole if you remain absent from its occasion prospect — consider of it like the gravitational arena of a planet. This sector is the point of no reappearance, when you are too near for some expectation of release. But you can securely detect the black hole from external of this arena. By allowance, this denotes its probable unbearable for a black hole to gulp up all in the Universe.

Fact 10: Black holes are utilized all the period in science narrative.

There are so lots of movies and cinemas using black holes, for instance, that it's unbearable to list them all. A stellar trip by the universe comprises a photograph look at a black hole. Occasion Prospect discovers the wonder of synthetic black holes — somewhat that is as well deliberated in the Star Trek universe. Black holes are also speaking regarding in: Galactica, Battlestar, Stargate: SG1 and lots of, numerous further space demonstrations.

Here on Universe Nowadays we have an excessive object regarding an applied usage for black holes— as space lab machines. No one can acquire to a black hole deprived of space tourism. Astronomy cast suggestions a good incident regarding stellar tourism.

7

Nebula

Introduction

Nebula is an actually amazing object to see. Called afterward the Latin word for "cloud" Nebula is not just enormous clouds of hydrogen, helium gas and plasma, and dust; they are as well frequently cosmological nurseries – *i.e.* the place where stars are born. And for centuries, far galaxies were frequently wrong for these enormous clouds.

Unfortunately, such explanations just scratch the exterior of what Nebula is and what their implication is. Among their development procedure, their functions in planetary and stellar creation and their assortment, Nebula have providing mortality with boundless plotting and innovation. For a few time now, researchers and astrophysicists have been conscious that external space is not actually a complete void. Actually, it is created up of gas and dust elements recognized cooperatively as the interstellar average. Around 99% of the Interstellar Medium is collected of gas, whereas around 75% of its quantity gets the form of hydrogen and the residual 25% as helium.

The interstellar gas contains partially of neutral particles and molecules, with charged atoms, such as electrons and ions. This gas is very thin, with a normal thickness of around 1 particle per cubic centimeter. In difference, Earth's environment has a thickness of about 30 quintillion particles per cubic centimeter at ocean level. Even although the interstellar gas is extremely detached, the quantity of substance enhances up above the massive aloofness among the stars. And finally, and with sufficient gravitational magnetism among clouds, this substance can merge and failure to customs planetary and stars systems.

Nebula Creation

In spirit, a nebula is designed when lots of the interstellar medium experience gravitational breakdown. Communal gravitational magnetism reasons substance to mass collected, creating sections of larger and larger thickness. From this, stars can form in the focus of the failing substantial, whose electromagnetic ionizing emission reasons the nearby gas to converted observable at visual frequencies. Maximum Nebula are massive in size, determining up to hundreds of light years in width. Though thicker than the space nearby them, most Nebulas are distant fewer thick than any void produced in an earthen atmosphere. Actually, a nebular cloud that was same in size to Earth would only so much substantial that its form would be only some kilograms.

Category of Nebula

Stellar substances that can be known as Nebula arise in four main types. Most reduction into the type of Diffuse Nebula, which donates they have no precise limits. These can be sectioned into two more types depend on their behavior with observable light – Radiation Nebula and Replication Nebula. Radiation Nebula are those that produce spectral line emission from ionized gas, and are frequently known as HII areas as they are mostly collected of ionized hydrogen. In difference, Replication Nebula do not produce important quantities of observable light, but are still shining as they reproduce the light from adjacent stars.

This is also called Dark Nebula, impervious clouds that do not produce observable emission and are not lightened by stars, but block light from shining substances after them. Much like Radiation and Replication Nebula, Dark Nebula are causes of electromagnetic radiations, mainly because of the attendance of dust in them.

A few Nebulas are designed as the consequence of supernova detonations, and are hence categorized as a Supernova Remnant Nebula. In this case, short-lived stars capability collapse in their centers and shock off their exterior coatings. This detonation leaves after a residue in the form of a solid thing – *i.e.* a neutron star and a cloud of dust and gas that is ionized by the power of the detonation. Further Nebula can create as Planetary Nebula, which comprises a short-form star arrive the last level of its life. In this situation, stars arrive to their Red Giant stage, sluggishly dropping their external coatings because of helium sparks in their internal. When the star has lost sufficient substantial, its heat rises and the UV radiation it produces ionizes the close substantial it has thrown off. This type also comprises the sub type called Proto planetary Nebula, which applies to astronomical substances that are suffering a short-lived occurrence in a star's growth. This is the fast stage that gets place among the Late Asymptotic Giant Branch and the subsequent Planetary Nebula stage.

In the Asymptotic Giant Branch stage, the star experiences form loss, producing a circum stellar shell of hydrogen gas. When this stage derives to a conclusion, the star arrive the Proto planetary Nebula stage, where it is animated by a dominant star, causing it to produce solid electromagnetic emission and converted a replication nebula. The Proto planetary Nebula stage remains until the dominant star spreads a heat of 30,000 K, later which it is hot sufficient to ionize the adjacent gas.

Past of Nebula Observance

Lots of nebulas substances were observed in the night sky by astrophysicists in Standard Ancient times and the Mid Ages. The first verified observance took place in 150 CE, when Ptolemy renowned the occurrence of five stars in Almagest that seemed nebulous in his book, he also renowned an area of light among the collections Leo and Ursa Major that was not related with any noticeable star.

On 4th July, 1054, the supernova that produced the Crab Nebula was observable to astrophysicists on Earth, and verified explanations that were created by both Chinese and Arabic astrophysicists have been recognized. While subjective sign occurs that other people observed the supernova, no records have been exposed. By the 17th century, developments in telescopes led to the first established explanations of Nebula. This initiated in 1610, when French astrophysicist Nicolas-Claude Fabri de Peiresc created the first detailed observance of the Orion Nebula. In 1618, Swiss astrophysicist Johann Baptist Cysat also experiential the nebula; and by 1659, Christian Huygens created the first comprehensive learning of it.

In the 18th century, the amount of experiential Nebula initiated to rise and astrophysicists initiated to accumulate lists. In 1715, Edmund Halley distributed a list of six Nebula – M11, M13, M22, M31, M42, and the Omega Centauri spherical group– in his An version of some Nebula or articulate spots like clouds, currently exposed between the fix stars by aid of the telescope. In 1746, French astrophysicist Jean-Philippe de Cheseaux accumulated a list of 20 Nebula, comprised eight that were not before recognized. Among 1751 and 53, Nicolas Louis de Lacaille classified 42 Nebula from the cape of good expectation, maximum of which were initially unidentified. And in 1781, Charles Messier accumulated his set of 103 Nebula, although a few were comets and galaxies.

The amount of experiential and classified Nebula significantly extended thanks to the exertions of William Herschel and his sister, Caroline. In 1786, the two distributed their list of One Thousand New Nebula and Groups of Stars, which was tracked up in 1786 and 1802 by a second and third set. At the time, Herschel supposed that these Nebula were only vague groups of stars, a faith he would adjust in 1790 when he experiential a factual nebula adjacent an aloof star. Opening in 1864, English astrophysicist William Huggins initiated to distinguish Nebula depend on their ranges. Unevenly one-third of them had the radiation range of a gas whereas the rest presented a constant range, reliable with a form of stars.

In 1912, American astrophysicist Vesto Slipher connected the subgroup of Replication Nebula after perceiving how a nebula adjacent a star coordinated the ranges of the Pleiades open group. By 1922, and as portion of the excessive discussion regarding the nature of curved Nebula and the mass of the universe, it had convert clear that lots of the earlier experiential Nebula were in detail aloof curved galaxies. In that similar year, Edwin Hubble declared that closely all Nebulas are connected with stars and that their brightness arises from star light. Since that time, the amount of correct Nebula has developed substantially, and their arrangement has been sophisticated thanks to developments in experimental apparatus and spectroscopy. In a word, Nebula is not only the initial points of stellar development, but can also be the termination point. And among all the star structures that fill our universe and our galaxy, nebulous clouds and multitudes are certain to be creating, only waiting to provide birth to the subsequent generation of stars.

8

Supernova

Introduction

Supernovae are explosion stars. They signify the extremely last phases of growth for a few stars. Supernova, as heavenly events are enormous discharges of marvelous power, as the star finishes occurring, with around 1020 times as much power created in the supernova detonation as our Sun discharges each second. Our Sun, providentially, will not end its lifecycle as a supernova. Presently, Supernova is only observed in galaxies other than the Cloudy Way. We recognize that Supernova have arisen in our Galaxy in the previous, since both Tycho Brahe and his protégé, Johannes Kepler, exposed bright Supernova arising in the Cloudy Way in 1572 and 1604, correspondingly. And, the Chinese, and others, have accounts of a visitor star arising in 1054 in the current collection Taurus. Nowadays, we realize leftovers of all three Supernova, which seem as increasing clouds of gas, where all was formerly exposed. Though, no supernova has been perceived in our Galaxy since Kepler's.

Supernova, when they are exposed, selected by the year in which they are exposed, and arrange in which they are exposed in that year, by with associates of the alphabet. For example, the fourth supernova exposed this last year was called SN 1998D, which arisen in the galaxy NGC 5440.

The sunniest supernova since Kepler's supernova was exposed on February 23, 1987, in the adjacent galaxy, the Large Magellanic Cloud. This supernova was simply perceived with the nude eye through 1987 in the Southern Hemisphere. This supernova was called SN 1987A. This supernova is still being experiential by an amount of telescopes, mainly, the Hubble Space Telescope. Additional bright current supernova, apparent from the Northern Hemisphere, was SN 1993J inside the galaxy Messier 81. As of 1998 January 1, 1270 Supernova have been exposed since Supernova initial really initiated to be classified in 1885, when a supernova went off in the adjacent Andromeda galaxy.

Kinds of Supernova

The presence of the spectrum permits astrophysicists to categorize Supernova into two major kinds—Category I and Category II. Fundamentally, Supernova rise from two extremely dissimilar types of stars huge ones and old, non-huge ones. The Category II Supernova is extremely powerfully demonstration the occurrence of the component hydrogen in their ranges. Category I Supernova does not demonstration some hydrogen in their ranges. The astrophysicist Rudolf Minkowski exposed this difference in 1941, and this category system was utilized for around five periods. It was supposed that Category II Supernova is the detonations of huge stars, while Category I Supernova rises from old, short-form stars.

In around 1985, things got a tiny extra complex. A few Category supernovas exposed and deliberate in the primary 1980s seemed to be odd in nature. They did not display a distinguishing spectral signature, supposed to be because of the occurrence of silicon, and perceived in a lot of other Category I Supernova ranges. Furthermore, some of these odd Supernovas displayed very strongly the presence of helium. Furthermore, these supernovas appeared to be occurring among populaces of huge stars in galaxies. For these causes, it was understood that Category I Supernova can be more sub categorized into those with the silicon spectral factor, and these were known as Category Ia Supernova, and those that do not demonstration this factor; this last collection were known as Category Ib Supernova.

Creation matters even extra complex, not all of the Category Ib Supernova since 1985 have presented the occurrence of helium in their ranges. These first colleagues of Category Ib Supernova are nowadays known as Category Ic Supernova. Extra and extra, supernova investigators have understood that the Category Ib/Category Ic difference includes excruciating hairs, and so, lots of such supernova experts put both of these Category I subsets into one major type: Category Ibc.

Supernova Arise

Supernova is realized to arise in galaxies all over the Universe. Galaxies are essentially categorized into three main sets —oval, irregulars, and spirals. Now, Category II and Category Ibc Supernova are realized to arise only in spiral and irregular galaxies, and these Supernovas also incline to be exposed in areas of these galaxies where star creation, mainly the creation of enormous stars, maximum surely has currently arisen in the last 10 million years or so. The Supernova has not been realized in elliptical galaxies. It is thus supposed that these Supernova rise from the detonations of enormous stars in galaxies.

Category IA Supernova is exposed in all three kinds of galaxies. But, Category IA Supernova is usually not initiate close enormous star creation. Since extremely tiny, if any, star creation arises nowadays in oval galaxies, it is supposed that Category Ia Supernova rise from elder, fewer enormous stars.

Philosophies of Supernova

In combination with this atmospherically indication for the nature of Supernova, astrophysicists, who grow physical philosophies to clarify heavenly wonders, and are thus usually known as philosophers by their associates, grow hypothetical replicas to clarify supernova detonations.

Nowadays, these replicas include cultured and composite computer replications of the detonations. What the philosophers incline to discovery is that stars extra enormous than around 8 solar multitudes, or, in other language, 8 times the form of our Sun, converted Category II and Category Ibc Supernova. These are young, comparatively enormous stars, which form in irregular and spiral galaxies. They also discovery that the Category Ia Supernova can finest be clarified by the detonation of rather unusual short-form stars called white dwarfs.

Stellar development is the learning of how stars change and alteration, both internally and externally, through their lives. Stars produce their personal power during their lives by the procedure of atomic fusion. The nuclei of sunnier components, such as helium and hydrogen, are enforced to fuse, or association, below the wonderful weights and hotness at or close the cores of stars, into the nuclei of weightier components. As Albert Einstein exposed, in his well-known mass-power correspondence value that everybody recognizes, $E=mc2$, power can be formed in big amounts from substance. When atomic responses arise intimate stars, these responses release enormous quantities of power, which inexorably dribbles out from the star's inner to its exterior, consequential in the light we perceive from the stars, their star shine.

In enormous stars, those extra enormous than around 8 solar multitudes, the order of nuclear fusion developments from the extremely easiest response of hydrogen nuclei to form helium nuclei, to extra complex responses, connecting the mixture, as it is recognized, of silicon nuclei into iron nuclei. The iron nucleus is the maximum constant nucleus in Nature, and it attacks fusing into any weightier nuclei, unless it is involuntary to do so with the effort of really difficult quantities of power. As a consequence, when the dominant core, as it is recognized, of a star converts clean iron nuclei, the central, which is usually the site of most of a star's power creation, is no lengthier capable to generate power and thus sustenance the star. The central can no lengthier sustenance the overwhelming power of gravity, consequential from all of the substance over the central, and the central thus failures below its personal heaviness.

.A few actually unusual physics gets place in this central failure. But, essentially, only neutrons can usually live the failure, and when the neutrons act composed below really unconceivable overwhelming weight to battle the failure, the core converts what is called a neutron star. The core then converts constant, but the respite of the enormous star is left in midpoint. The central failure quickly stops, and the central, like an embraced squeegee, bounds back, discharging a massive quantity of power, which rips by the external coatings of the star. The innovative enormous star dies in a burning detonation, with only the recently-designed neutron star living this massive detonation.

The star has finished its life as a Category II or Category Ibc supernova. And the expiry throes of this star arise very quickly, above only a time of numerous milliseconds! This associated to a star that, up to that point, had occurred for some million years. If the star initiated its life with an actually big quantity of form, the philosophers declare that not even the neutrons at the star's central can clutch back the overwhelming power of gravity. At that point, as the star finishes its life, the central converts a black hole. Probably, the consequence of the creation of the black hole is a supernova detonation, but some queries stay if this is actually the cable of actions for such extremely enormous stars.

Now, this type of development will not arise for the Sun. The Sun will remain to extremely silently anger its dominant hydrogen into helium for the subsequent five billion years or so. The central will develop clean helium, which will then fuse to carbon in a comparatively small time. Lastly, the carbon at the central cannot acquire hot sufficient to fuse into other kind of nucleus. The carbon central can no lengthier bear the Sun's power and failures below its individual heaviness, much as the extra composite centers of enormous stars do. Though, electrons in the central action to battle the failure, and the central of the Sun will converted what is called a white dwarf. As the creation of the core white dwarf arises, the external coatings of the Sun will be sloughed off into space to form a planetary nebula. As the nebula scatters above lots of thousands of years, the emaciated white dwarf remainder of the preceding Sun will sit in the Galaxy and glow away its remaining warmth above lots of billions of years.

White dwarfs, as you can suspicious, are not very enormous, since one will form from the central of the Sun, which nowadays comprises, by description, one solar form. In 1938 the Indian astrophysicist, S. Chandrasekhar, resolute that white dwarfs cannot be extra enormous in the cosmos than 1.4 solar masses. If a white dwarf were to beat this border, known as the Chandrasekhar limit, the star would stop to happen. So, if a white dwarf discovers itself in a binary star system, where the two stars are near sufficient that their communal gravity consequences in their collaboration, then the binary acquaintance can dump substance onto the white dwarf. The white dwarf's form gradually and progressively rises, to the point that it can surpass the Chandrasekhar border, if this occurs, then poof. The white dwarf detonates in a Category Ia supernova and is totally demolished. The substance that once was the white dwarf acquires cremated into harmful components, which deterioration above time, and remain to power the light curvature of the supernova.

Consequences of Supernova

When Supernova detonates, they have philosophical consequences on their environments in galaxies. The marvelous power that is unconventional moves the gas in its atmosphere, assertive on it and condensing it. If the gas was originally equally solid, then the flattened thicker gas can really go on to failure and usage new stars. The power of the detonation also manufactures new components, mainly those heavier than iron. These renewed, new components are then scattered into the adjacent vaporous moderate, elevating it. Thus, future generations of stars designed after the supernova comprises extra weighty components than earlier generations. Actually, the improvement of the gas in our area of the Cloudy Way stretched such a point that an enough amount of weighty components occurred to provide increase to lifetime, as we recognize it, here on Earth. Supernova are supposed to be directly liable for us everything.

Supernova also possible by minor nuclear and subatomic atoms out into the galaxies, which we call cosmic emissions, these atoms, affectingly the Cloudy Way Galaxy, pass by space and impose on the Earth; it is supposed that these great-speed, great-power cosmic emissions can be partly accountable for inherited alteration and, thus, development of life here on Earth.

9

Dark Matters and Energy

Introduction

In the early 1990s, one object was equally convinced regarding the growth of the universe. It may have sufficient energy thickness to break its growth and recollapse, it may have so small energy thickness that it would never break increasing, but gravity was convinced to sluggish the growth as time went on. Decided, the decelerating had not been experimental, but, hypothetically, the universe had to sluggish. The universe is filled of matter and the fascinated power of gravity attractions all matter collected. Then derived 1998 and the Hubble space telescope explanations of extremely aloof supernovae that exhibited that, an extended time before, the universe was really increasing extra gradually than it is nowadays. So the growth of the universe has not been reducing because of gravity, as everybody supposed, it has been quickening. No one predictable this, no one recognized how to describe it. But something was affecting it.

Finally philosophers derived up with three kinds of clarifications. Possibly it was a consequence of an extended-rejected form of Einstein's philosophy of gravity, one that comprised what was known as a cosmological continuous. Possibly there was a few odd type of energy-liquid that occupied space. Possibly there is something incorrect with Einstein's philosophy of gravity and a new philosophy can comprise a few type of area that makes this interstellar quickening. Philosophers still don't recognize what the accurate clarification is, but they have specified the explanation a name. It is known as dark energy.

Dark Energy

Extra is unidentified than is identified. We recognize how much dark energy there is as we recognize how it influences the universe's growth. Further than that, it is a comprehensive anonymous. But it is a significant anonymous. It rotates out that approximately 68% of the universe is dark energy. Dark matter creates up around 27%. The rest - all on Earth, all forever experimental with all of our devices, all common matter - improves up to fewer than 5% of the universe. Come to consider of it, possibly it shouldn't be known as common matter at all, since it is such a minor portion of the universe.

One description for dark energy is that it is a thing of space. Albert Einstein was the first creature to understand that blank space is not nil. Space has wonderful things, lots of which are only start to be assumed. The first things that Einstein exposed are that it is conceivable for extra space to derive into life. Then one form of Einstein's gravity philosophy, the form that comprises a cosmological continuous, creates a second forecast— blank space can own it's possess energy. As this energy is material goods of space themselves, it would not be adulterated as space increases. As extra space arises into life, extra of this energy of space would look. As a consequence, this appear of energy would reason the universe to enlarge quicker and quicker. Inappropriately, no one recognizes why the cosmological continuous must even be there, greatly fewer why it would have accurately the correct value to cause the experimental hastening of the universe.

Alternative description for how space gets energy derives from the dramatic philosophy of matter. In this philosophy, unfilled space is really filled of provisional atoms that repeatedly appear and then evaporate. But when physicists annoyed to compute how much energy this would provide unfilled space, the response derived out incorrect - incorrect by many. The amount came out 10120 times too large. That's a 1 with 120 zeros afterward it. It's solid to acquire a response that bad. So the mystery stays.

Alternative description for dark energy is that it is a new type of energetically power liquid or arena, something that full all of space but something whose consequence on the growth of the universe is the conflicting of that of matter and common energy. A few philosophers have called this embodiment, after the fifth component of the Greek theorists. A last probability is that Einstein's philosophy of gravity is not accurate. That would not only distress the growth of the universe, but it would also distress the mode that common matter in galaxies and clusters of galaxies performed. This truth would give a mode to determine if the explanation to the dark energy difficulty is a new gravity philosophy or not: we can detect how galaxies come composed in groups. But if it does go out that a new philosophy of gravity is desirable. The thing that is required to determine among dark energy potentials - a property of space, a new active liquid, or a new philosophy of gravity - is extra statistics, improved statistics.

Dark Matter

By appropriate a hypothetical classical of the configuration of the universe to the joint set of cosmological explanations, researchers have derive up with the configuration that we defined beyond, ~68% dark energy, ~27% dark matter, ~5% common matter.

We are much extra convinced what dark matter is not than we are what it is. Initially, it is dark, denotation that it is not in the appearance of planets and stars that we perceive. Explanations express that there is faraway too tiny observable matter in the universe to structure the 27% necessary by the explanations. Next, it is not in the appearance of dark clouds of common matter, matter created up of atoms known as baryons. We recognize this as we would be capable to identify baryonic clouds by their captivation of emission transitory by them. Third, dark matter is not antimatter, as we do not realize the exclusive gamma emissions that are created when antimatter defeats with matter. Lastly, we can regulation out big galaxy-sized black holes on the basis of how lots of gravitational lenses we get. High attentions of matter curve light transitory close them from matter more away, but we do not realize sufficient lensing actions to recommend that such matter to create up the necessary 25% dark matter influence.

Though, at the present, there are still some dark matter potentials that are feasible. Baryonic matter can still structure the dark matter if it were all tense up in brown dwarfs or in minor, thick masses of weighty components. These potentials are called enormous compressed halo matters, or MACHOs. But the maximum general view is that dark matter is not baryonic at all, but that it is created up of other, extra striking atoms like axioms or weakly interrelating enormous atoms.

Influences of Dark Matter

The dark matter difficulty can also be observed as a query of the nature of gathering matter. Dark matter should be the fundamental element of the biggest constructions in the universe—clusters and galaxies. Deprived of dark matter, the universe would be an extremely dissimilar place, as per recent philosophies.

And dark matter is not just for explanation the behavior of aloof bodies in the cosmos, but is copious in our galaxy too. It is predictable that our solar system is transitory by a fine ocean of dark matter atoms with a thickness as great as unevenly 105 per cubic meter. We can expectation to identify the mutability of dark matter transitory by the Earth, and even to identify the terms of dark matter, conforming to the times of year when the Earth is affecting with, or alongside, the flow of dark matter orbiting the core of the Cloudy Way.

Influences of Dark Energy

It is liable for the cosmic moving, and worldwide groups of astrophysicists are occupied to improve dimensions of that quickening. At stake is decision on Einstein's highest blooper, potential insight into the important philosophy of nature, and the destiny of the universe.

It is attractive to attempt to join the clarifications for dark matter and dark energy, but there are excessive dissimilarities among the two, dark matter appeals and dark energy impulses. That is, dark matter is appealed to clarify better-than-predictable gravitational magnetism. In difference, dark energy is raised to clarify feebler-than-predictable, and actually negative, gravitational magnetism. Additionally, the consequences of dark matter are obvious on extent balances unevenly 10 super parsecs and minor, while dark energy seems only to be applicable on balances of unevenly 1,000 super parsecs or better. Lastly, it is significant to query whether the dark matter and dark energy wonders can have gravitational clarifications. Possibly the rules of gravitation fluctuate from Einstein's philosophy. This is surely a probability, but so far common irrelativeness has not unsuccessful a sole experiment. And arresting new visions of collections have exposed behavior that is unpredictable with a gravitational cure-- meaning that dark matter truly is there.

10

Stars

Introduction

Stars are the maximum broadly known astronomical substances, and signify the maximum essential elements of galaxies. The distribution, composition, and age of the stars in a galaxy follow the past, subtleties, and development of that galaxy. Furthermore, stars are liable for the production and dissemination of weighty components such as nitrogen, oxygen, and carbon, and their features are closely tied to the features of the planetary structures that can combine regarding them. Thus, the learning of the birth, lifecycle, and expiry of stars is core to the area of astronomy.

A star is an enormous scope of extremely hot, shining gas. Stars create their possess energy and light by a procedure known as nuclear fusion. Fusion occurs when lighter components are required to converted weightier components. When this occurs, a wonderful quantity of energy is produced causing the star to warmth up and sparkle. Stars derive in a range of dimensions and colors. Our Sun is normal extend yellowish star, stars which are minor than our larger stars are blue and Sun are reddish.

Star Creation

Stars are born in the clouds of dust and dispersed through maximum galaxies. A well-known instance of such as a dust cloud is the Orion Nebula. Commotion profound in these clouds provide increase to bulges with enough form that the dust and gas can initiate to failures below its personal gravitational magnetism. As the cloud failures, the substantial at the centre initiates to warmth up. Called a proto star, it is this warm central at the heart of the failing cloud that will one day developed a star. 3D computer replicas of star creation forecast that the rotating clouds of failing dust and gas can splitting up into two or three blobs; this would clarify why the mainstream the stars in the Cloudy Way are combined or in collections of numerous stars.

As the cloud failures, a thick, warm centre appearance and initiates assembly gas and dust. Not all of this substantial ends up as portion of a star — the residual dust can developed asteroids, comets, or planets or can stay as dust.

Origin of Stars

Stars are originates by extremely hot gases. These gases are generally helium and hydrogen, which are the two brightest components. Stars sparkle by fiery hydrogen into helium in their centers, and advanced in exists made weightier components. Maximum stars have minor quantities of weightier components like iron, oxygen, nitrogen and carbon, which were produced by stars that occurred previously them. Afterward a star turns out of fuel; it refuses greatly of its substantial rear into space. New stars are made from this substantial. So the substantial in stars is used.

You may not be amazed to recognize that stars are created of the similar material as the rest of the Universe —25% helium,73% hydrogen, and the final 2% is all the other components. That's it. Excluding for some dissimilarities here and there, stars are created of pretty much the similar material. Afterward the Big Bang, 13.7 billion years before; the whole Universe was a warm thick range. The circumstances in this early Universe were so warm that it was equal to being in the centre of a star. In other languages, the whole Universe was like a star.

The Universe reserved growing and chilling down, and finally the helium and hydrogen chilled down to the point that it can really start gathering collected with its common gravity. This is how the earliest stars were born. And like the stars we have nowadays, they were created up of unevenly25% helium and 73% hydrogen. These first stars were massive and possibly exploded as supernovae in a million years of creating, in their lifetime, and in their expiry, these first stars produced few of the weightier components that we have here on Earth, such as gold, carbon, uranium and oxygen.

Stars have been creating since the Universe initiated. Actually, astrophysicists compute that 5 new stars form in the Cloudy Way each year. Some have extra of the weightier components left above from earlier stars; these are metallic-rich stars. Others have fewer of these components; the metallic-poor stars. But even so, the proportion of components is still unevenly the similar. Our own Sun is an instance of a metallic rich star, with a greater than normal quantity of weightier components in it. And yet, the Sun's proportions are extremely same — 27.1% helium, 71% hydrogen, and then the rest as weightier components, such as carbon, nitrogen, oxygen, etc. Obviously, the Sun has been changing hydrogen into helium in its centre for 4.5 billion years.

Stars everywhere are created of the similar material —1/4 helium and 3/4 hydrogen. It's the material left above from the creation of the Universe, and one of the maximum sophisticated portions of signal to aid clarify how we are here nowadays.

Dissimilar Kinds of Stars

All things considered, there are lots of dissimilar kinds of stars, ranging from small brown dwarfs to red and blue supergiant. There are even extra strange types of stars, like stars Wolf-Rayet and neutron stars. And as our investigation of the Universe remains, we remain to study things regarding stars that power us to increase on the mode we consider of them. Let's get an appearance at all the dissimilar kinds of stars there are.

Protostar

A protostar is what you have previously a star mass. A protostar is a group of gas that has distorted down from a huge molecular cloud. The protostar stage of stellar growth lasts around 100,000 years. Above time, gravity and weight rise, compelling the protostar to failure down. All of the energy relief by the protostar derives only from the warming caused by the gravitational energy – nuclear combination responses have-not happening yet.

T Tauri Star

A T Tauri star is phase in a star's creation and growth right previously it converts a major system star. This stage arises at the end of the protostar stage, when the gravitational weight holding the star collected is the basis of completely its energy. T Tauri stars don't have sufficient weight and hotness at their centers to produce nuclear combination, but they do be like major system stars; they are regarding the similar heat but sunnier as they are a bigger. T Tauri stars can have big regions of resort attention, and have penetrating X-ray flashes and very influential stellar winds. Stars will stay in the T Tauri phase for around 100 million years.

Major System Star

The mainstream of all stars in our galaxy, and even the Universe, are major system stars. Our Sun is a major system star, and so are our adjacent neighbors, Alpha Centauri A and Sirius. Major system stars can differ in dimension, size and intensity, but they are all doing the similar thing — changing hydrogen into helium in their centers, discharging a wonderful quantity of energy. A star in the major system is in a state of hydrostatic stability. Gravity is dragging the star inner and the light weight from all the combination responses in the star are pushing outer. The inner and outer powers equilibrium one added out, and the star preserves a round shape. Stars in the major system will have a dimension that based on their size, which describes the quantity of gravity dragging them inner.

The inferior size boundary for a major system star is around 0.08 times the size of the Sun, or 80 times the size of Jupiter. This is the smallest quantity of gravitational weight you require to burn combination in the centre. Stars can hypothetically produce to extra than 100 times the size of the Sun.

Red Giant Star

When a star has expended its stock of hydrogen in its centre, combination ends and the star no lengthier produces an external weight to counter the inner weight dragging it collected. A shell of hydrogen nearby the centre burns with the life of the star, but causes it to rise in size intensely. The old star has developed a red giant star, and can be 100 times bigger than it was in its major arrangement stage. When this hydrogen fuel is utilized up, additional shells of helium and even weightier components can be expended in combination responses. The red giant stage of a star's life will only last some hundred million years ago it runs out of fuel totally and converts a white dwarf.

White Dwarf Star

When a star has totally run out of hydrogen fuel in its central and it deficiencies the form to power greater components into combination response, it develops a white dwarf star. The external light weight from the combination response breaks and the star failures inner below its personal gravity, a white dwarf sparkles as it was a warm star once, but there is no combination responses occurring any longer. A white dwarf will just cool down until it as the contextual heat of the Universe. This procedure will get hundreds of billions of years, so no white dwarfs have really chilled down that distant yet.

Red Dwarf Star

Red dwarf stars are the maximum general type of stars in the Universe. These are major system stars but they have such short form that they are much chiller than stars like our Sun. They have additional benefit. Red dwarf stars are capable to stay the hydrogen fuel combining into their central, and so they can preserve their fuel for much lengthier than other stars. Astrophysicist's evaluation that some red dwarf stars will scorch for up to 10 trillion years, the lowest red dwarfs are 0.075 times the amount of the Sun, and they can have a form of up to half of the Sun.

Neutron Stars

If a star has among 1.35 and 2.1 times the form of the Sun, it doesn't appear a white dwarf when it expires. In its place, the star expires in a disastrous supernova detonation, and the residual central converts a neutron star. As its name indicates, a neutron star is an unusual kind of star that is collected completely of neutrons. This is because the penetrating gravity of the neutron star infatuations protons and electrons collected to appear neutrons. If stars are even extra enormous, they will develop black holes as a replacement for of neutron stars after the supernova goes off.

Supergiant Stars

The biggest stars in the Universe are supergiant stars. These are monsters with loads of times the form of the Sun. Dissimilar a comparatively constant star like the Sun, supergiant are overwhelming hydrogen fuel at an massive rate and will consume all the fuel in their centres in just some million years. Supergiant stars live fast and die young, exploding as supernovae; totally decomposing themselves in the procedure. As you can realise, stars arise in a lot of masses, ranges and colours. Significant what explanations for this, and what their different life periods appearance like, are all significant when it derives to sympathetic our Universe. It also aids when it arises to our continuing determinations to discover our local stellar area, not to reference in the search for interplanetary life.

11

The Moon

Introduction

The moon is the simplest heavenly thing to get in the night sky — when it's there. Earth's only natural satellite floats beyond us bright and curved until it apparently evaporates for some nights. The regularity of the moon's stages has directed mortality for periods — for example, calendar months are unevenly equivalent to the time it gets to go from one complete moon to the next.

Moon stages and the moon's orbit are obscurities to a lot of. For instance, the moon continuously shows us the similar face. That occurs as it gets 27.3 days both to swap on its axis and to orbit Earth. We get both the full moon; half-moon and new moon as the moon replicates sunlight. How much of it we see based on the moon's location in relative to sun and the Earth. Although a satellite of moon, the Earth, with a width of around 2,159 miles, is larger than Pluto. The moon is a bit extra than one-fourth the mass of Earth, a much slighter ratio than some other planets and their moons. This denotes the moon has a great consequence on the planet and extremely probably is what creates life on Earth conceivable.

The Moon Appearances

There are different philosophies regarding how the moon was formed, but current evidence specifies it designed when an enormous crash ripped a mass of Earth away.

The foremost clarification for how the moon designed was that a huge effect bashed off the rare elements for the moon off the embryonic liquefied Earth and into orbit. Researchers have recommended the effect was unevenly 10 percent the size of Earth, around the mass of Mars. As Earth and the moon are so parallel in configuration, scientists have determined that the effect should have arisen around 95 million years after the creation of the solar system.

Though the big effect philosophy leads the technical group's conversation, additional philosophy recommends that two young moons can have hit to appear a sole big one. Earth can even have stolen the moon from Venus, as per a current philosophy.

Inner Construction

The moon extremely probable has an extremely minor center just 1 to 2 percent of the moon's size and unevenly 420 miles extensive. It probable contains generally of iron, but canas well comprise big quantities of sulfur and other components.

Its stony layer is around 825 miles dense and created up of thick rocks rich in magnesium and iron. Lavas in the layer created their mode to the outward in the historical and exploded volcanically for extra than a billion years — from as a minimum four billion years before to less than three billion years past. The coating on top medians a few 42 miles deep, the outmost portion of the coating is jumbled and cracked because of all the big effects it has established, a devastated region that provides approach to complete substantial under a deepness of around 6 miles.

Surface Configuration

Like the four internal planets, the moon is stony. It's blemished with hollows designed by asteroid effects millions of years before. As there is no climate, the hollows have not battered. The normal configuration of the lunar surface by mass is roughly 3 percent aluminum, 10 percent iron,19 percent magnesium,20 percent silicon,3 percent calcium, 43 percent oxygen, 0.18 percent titanium,0.12 percent manganese and 0.42 percent chromium.

Orbiters have created drops of water on the lunar surface that can have invented from profound subversive. They have also situated hundreds of depths that can house voyagers who stay on the moon long-time.

Environment of the Moon

The moon has an extremely slim environment, so a coating of dust — or a footmark — can sit uninterrupted for centuries. And deprived of much of an environment, warmth is not detained close the outward, so heats differ eagerly. Daytime heats on the sunny side of the moon extent 273 grades F (134 C); on the dark side it acquires as icy as detriment 243 F (minus 153 C).

Orbital Features

- Perigee (nearby approach to Earth) — 225,700 miles.
- Apogee (farthest aloofness from Earth) — 252,000 miles.
- Normal distance from Earth— 238,855 miles.

Earth/Orbit Connection

The moon's gravity pulls at the Earth, affecting foreseeable increases and drops in ocean stages called tides, to a greatly minor amount, tides as well arise in ponds, the environment, and in Earth's layer. Great tides are when water protuberances mounting, and little tides are when water falls down. Great tide consequences on the adjacent of the Earth close the moon because of gravity, and it as well occurs on the side furthest from the moon because of the apathy of water. Little tides arise among these two bulges.

The pull of the moon is also reducing the Earth's revolution, and consequence called tidal slowing, which raises the extent of our day by 2.3 milliseconds per century. The energy that Earth drops is get up by the moon, growing its distance from the Earth, which denotes the moon acquires beyond away by 3.8 centimeters yearly.

The moon's gravitational pull can have been enter to creating Earth a livable planet by curbing the amount of vibrates in Earth's axial slope, which led to a comparatively constant weather above billions of years where life can embellishment. The moon doesn't leakage from the interaction unharmed. A new learning recommends that Earth's gravity strained the moon into its strange form primary in its lifetime.

Lunar Eclipses

In eclipses, the moon, sun and Earth are in a straight line, or closely so. A lunar eclipse gets position when Earth acquires directly or nearly directly among the moon and the sun, and Earth's shadow drops on the moon. A lunar eclipse can arise only through a full moon. A solar eclipse arises when the moon acquires directly or almost directly among the Earth and sun, and the moon's shadow drops on us. A solar eclipse can arise only in a new moon.

Investigation

Some antique persons supposed the moon was a bowl of fire, whereas others supposed it was a mirror that replicated Earth's oceans and lands, but antique Greek theorists recognized the moon was a sphere circling the Earth whose moon shine replicated sunlight. The Greeks also supposed the dark regions of the moon were oceans whereas the bright areas were land, which prejudiced the present terms for those places — "terrae" and "maria," which is Latin for oceans and land, correspondingly.

The innovative astrophysicist Galileo Galilei was the first to usage a telescope to create technical explanations of the moon, determining in 1609 an uneven, mountainous area that was rather dissimilar from the common views of his day that the moon was even.

Thank you for reading!

Education is something extremely passionate about fact so read books online for education, knowledge flow offer a wide variety of topics like engineering, science, business & management and school books for student and learners. You can find more than 100+ popular education books at best price in India on Knowledge flow. Read books online from huge selection of learning resources & education books.

www.ingramcontent.com/pod-product-compliance
Lightning Source LLC
Chambersburg PA
CBHW070812220526
45466CB00002B/644